Humpty Dumpty

The enormous turnip

- Can sequence events. Notes/date:

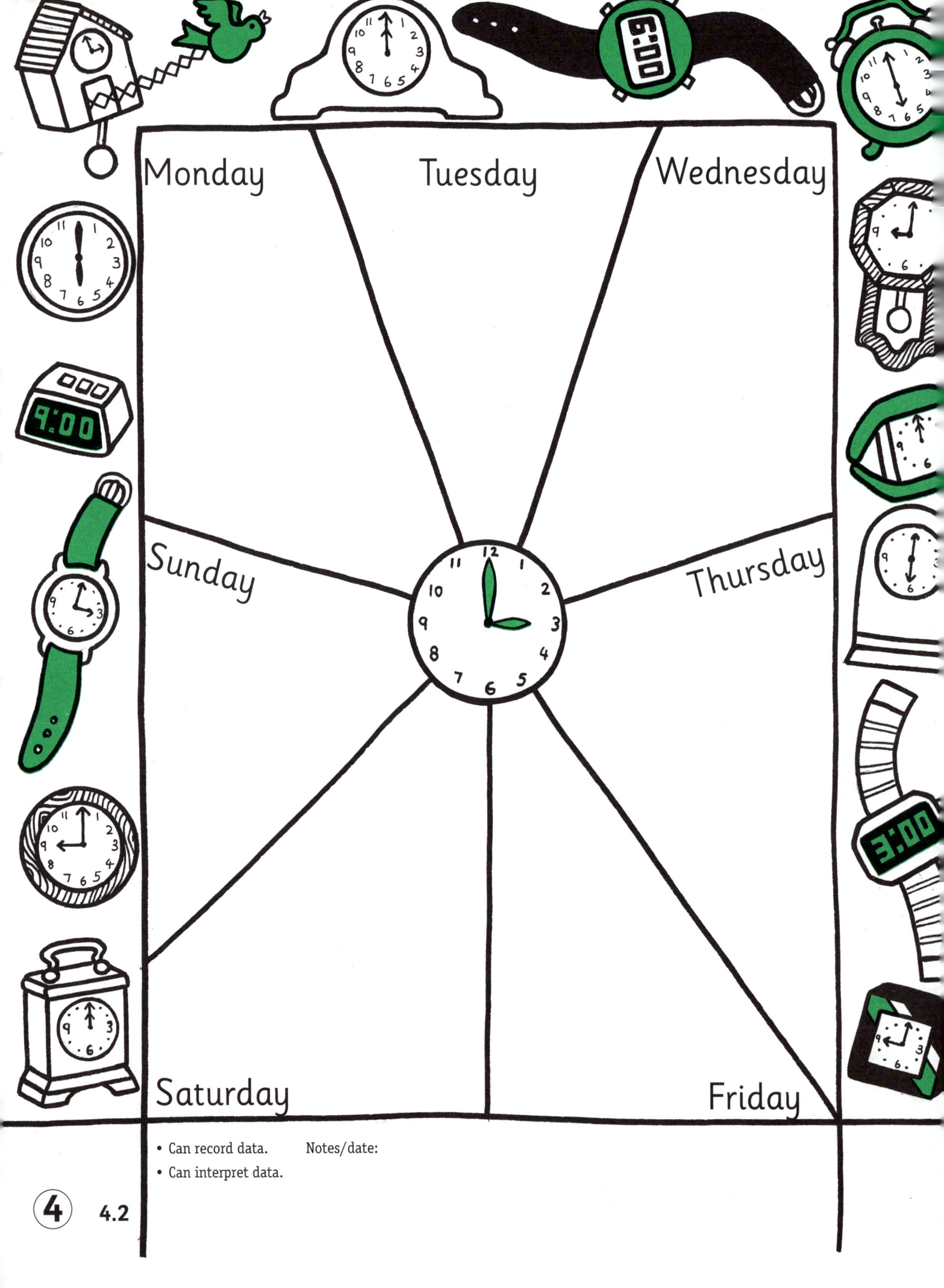

- Can record data.
- Can interpret data.

Notes/date:

4.2

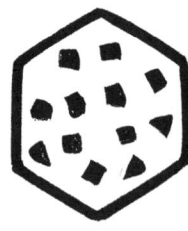
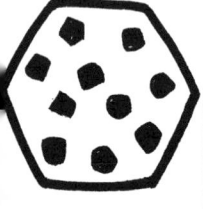

What did you do with your cookies?
Write and draw.

- Understands number bonds to 5. Notes/date:
- Can record in own way.

4.3

| 1 | 2 | 3 | 4 | 5 | 6 | 7 | 8 | 9 | 10 |

5 hop back 3, land on 2

| 1 | 2 | 3 | 4 | 5 | 6 | 7 | 8 | 9 | 10 |

4 hop back 1, land on 3

| 1 | 2 | 3 | 4 | 5 | 6 | 7 | 8 | 9 | 10 |

4 hop back 2, land on

| 1 | 2 | 3 | 4 | 5 | 6 | 7 | 8 | 9 | 10 |

__ hop back __ land on __

| 1 | 2 | 3 | 4 | 5 | 6 | 7 | 8 | 9 | 10 | 11 | 12 | 13 | 14 | 15 | 16 |

__ hop back __ land on __

• Can count back.   Notes/date:

4.4

6 bees
cross out 3 bees
leaves 3

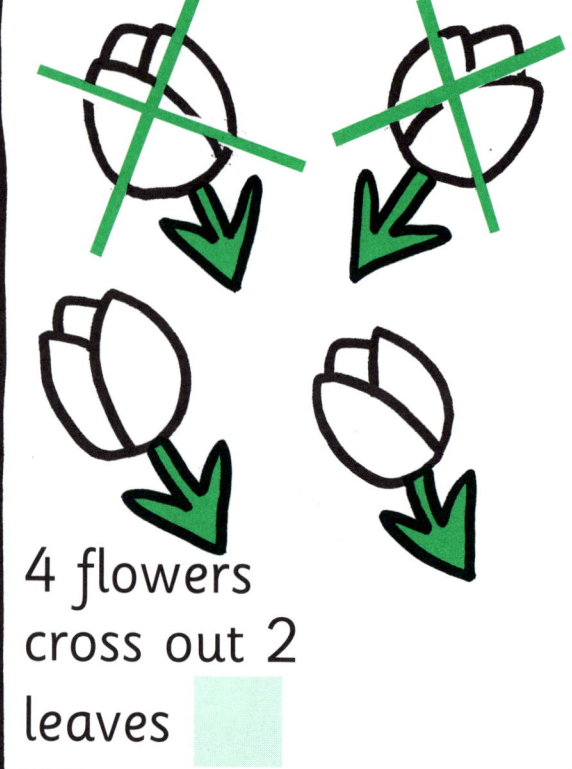

4 flowers
cross out 2
leaves

5 snakes
cross out 3
leaves

Draw

3 cross out 2
leaves

- Can 'take away', hop back, cross out.
- Can talk about work.

Notes/date:

4.4

5 cross out 3
leaves 2.
5 − 3 = 2

7 cross out 4
leaves 3.
7 − 4 =

8 cross out 4
leaves 4.
8 − 4 =

10 cross out 5
leaves 5.
10 − 5 =

Draw a crossing out picture.

____ cross out ____

leaves ____

- Can talk about subtraction.   Notes/date:
- Can use calculator.

4.4

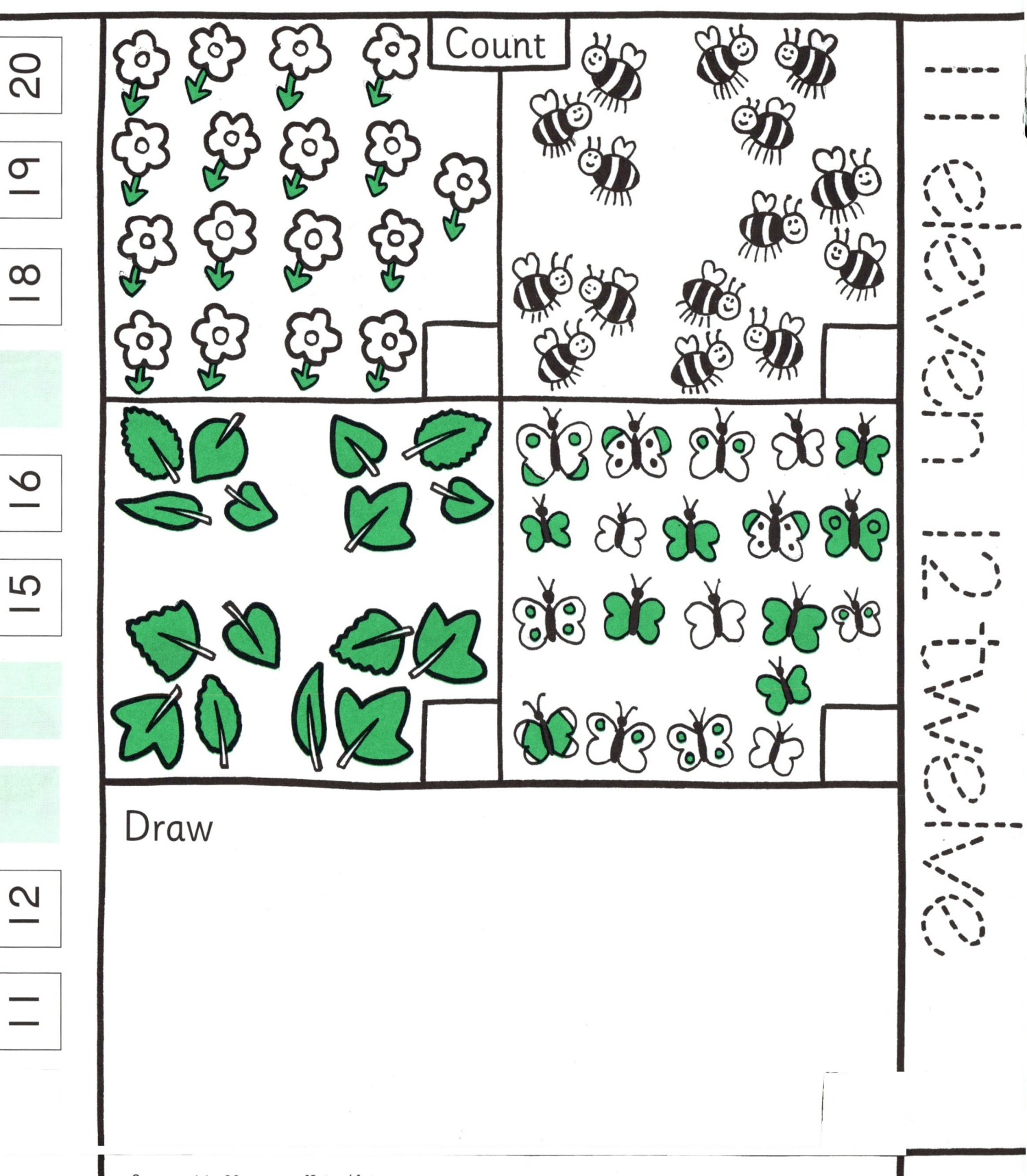

Snakes

Make a short snake a long snake

Measure them. Use  string cubes

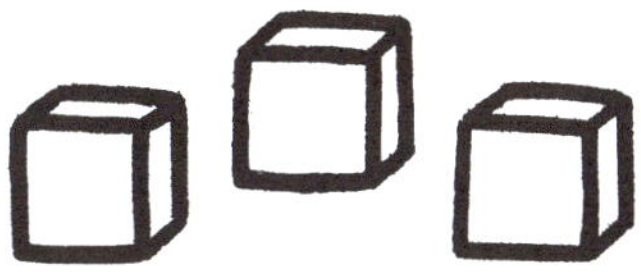

short snake

long snake

___ cubes long   ___ cubes long

- Understands long/short.
- Can use language of linear measure.

Notes/date:

P.B9

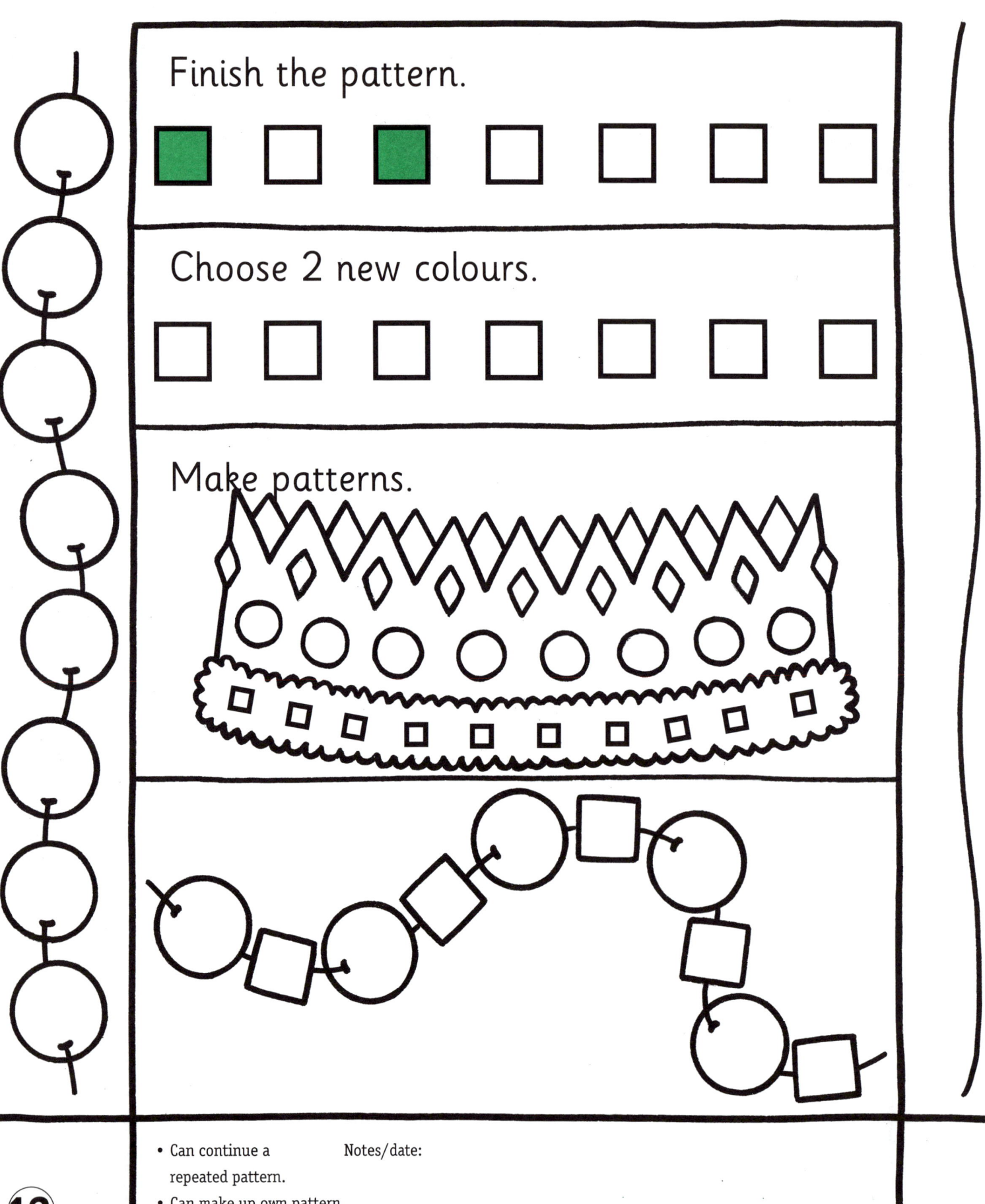

Finish the pattern.

Choose 2 new colours.

Make patterns.

- Can continue a repeated pattern.
- Can make up own pattern.

Notes/date:

- Can recognise and name a star.
- Can use appropriate language.

Notes/date:

4.7

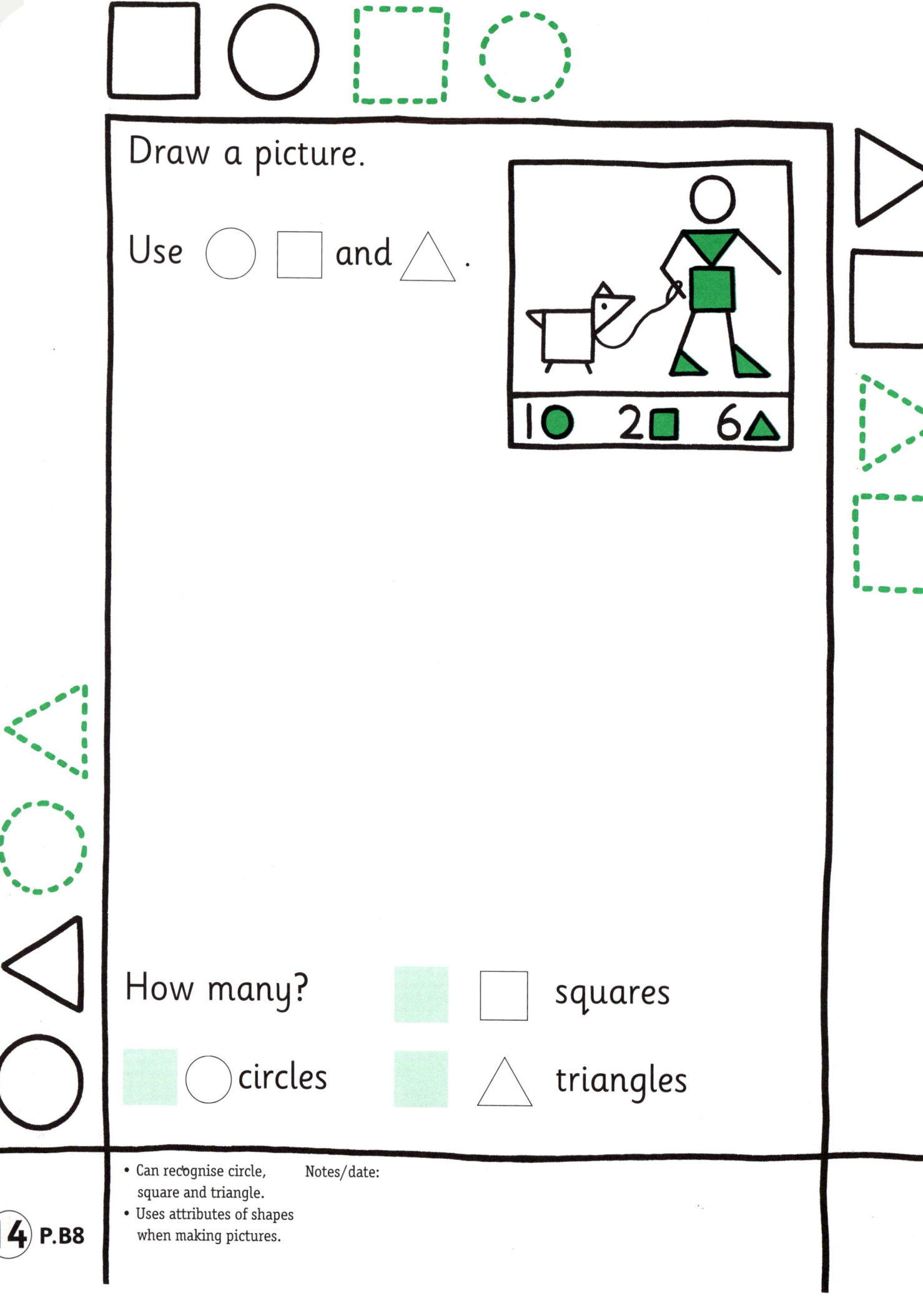